5-95.

THE OTHER HOUSE

The Other House

ANNE STEVENSON

Oxford   New York

OXFORD UNIVERSITY PRESS

1990

Oxford University Press, Walton Street, Oxford OX2 6DP

Oxford New York Toronto
Delhi Bombay Calcutta Madras Karachi
Petaling Jaya Singapore Hong Kong Tokyo
Nairobi Dar es Salaam Cape Town
Melbourne Auckland
and associated companies in
Berlin Ibadan

Oxford is a trade mark of Oxford University Press

First published 1990 as an Oxford University Press paperback

British Library Cataloguing in Publication Data
Stevenson, Anne, 1933–
The other house. – (Oxford poets)
I. Title
821.914
ISBN 0–19–282739–1

Library of Congress Cataloging in Publication Data
Stevenson, Anne, 1933 Jan. 3–
The other house / Anne Stevenson.
p. cm.
I. Title.
PR6069.T4508 1990 821'.914—dc20 89–49305
ISBN 0–19–282739–1

Typeset by Wyvern Typesetting Ltd
Printed in Great Britain by
J. W. Arrowsmith Ltd, Bristol

For my grandson Paul
and his parents

Houses we've lived in
inhabit us
and history's restless
in the rooms of the mind.
　　　　　—*Gillian Clarke*

Acknowledgements

Poems from this collection have appeared in the *Times Literary Supplement*, *P N Review*, *Poetry Review*, *Poetry Ireland Review*, *Partisan Review*, the *Atlantic*, *Graham House Review*, the *New Welsh Review*, the *Sunday Times*, the *Observer*, *Lines Review*, and the *Green Book*.

'Calendar', 'North Easter', 'Night Walking with Shadows', and 'The Morden Angel' were included in my pamphlet *Winter Time*, published by MidNAG in 1986. Two of the Seven Poems after Francis Bacon appeared in *With a Poet's Eye*, Tate Gallery, 1986. Three poems were first published in Poetry Book Society Anthologies: 'Study for a Portrait of Van Gogh' in 1985/86; 'What I miss' in 1987/88, and 'Eros' in 1988/89. 'And even then' was written for the anthology *Tidelines* (1988) to support the Druridge Bay anti-nuclear power campaign.

Special and warm thanks are due to the Scottish Arts Council and to the University of Edinburgh for giving me time, money, and a room in the David Hume Tower, where I wrote many of these poems between October 1987 and June 1989.

ANNE STEVENSON
October 1989

Contents

In the Nursery

I lift the seven months baby from his crib,
a clump of roots.
Sleep drops off him like soil
in clods that smell sunbaked and rich with urine.
He opens his eyes,
two light blue corollas.
His cheek against mine
is the first soft day in the garden.
His mouth makes a bud, then a petal,
then a leaf.
In less than seven seconds
he's blossoming in a bowl of arms.

The Other House

In the house of childhood
I looked up to my mother's face;
the sturdy roofbeam of her smile
buckled the rooms in place.
A shape of the unchangeable
  taught me the word 'gone'.

In the house of growing up
I lined my prison wall
with lives I worshipped as I read.
If I chose one, I chose all,
such paper clothes I coveted
  and ached to try on.

The house of youth has a grand door,
a ruin the other side
where death watch & company
compete with groom and bride.
Nothing was what seemed to be
  in that charged dawn.

They advertised the house of love,
I bought the house of pain,
with shabby little wrongs and rights
where beams should have been.
How could those twisted splintered nights
  stand up alone?

My angry house was a word house,
a city of the brain,
where buried heads and salt gods
struggled to breathe again.
Into those echoing, sealed arcades
  I hurled a song.

It glowed with an electric pulse,
firing the sacred halls.
Bright reproductions of itself
travelled the glassy walls.
'Ignis fatuus', cried my voice.
    And I moved on.

I drove my mind to a strange house
infinitely huge and small,
a cone, to which a dew-drop earth
leeches, invisible.
Infinite steps of death and birth
    lead up and down.

Beneath me, infinitely deep,
solidity dissolves;
above me, infinitely wide,
galactic winter sprawls.
That house of the utterly outside
    became my home.

In it, the house of childhood
safeguards my mother's face.
A lifted eyebrow's 'Yes, and so?'
latches the rooms in place.
I tell my children all I know
    of the word 'gone'.

Talking Sense to my Senses

Old ears and eyes, so long my patient friends,
For you this silicon nerve and resin lens.
Guides when I heard and saw, yet deaf and blind
Stumbled astray in the mazes of my mind,
Let me assist you now I've lived to see
Far in the dark of what I have to be.

Shunted outside the hubbub of exchange,
Knowledge arrives, articulate and strange,
Voice without breath, light without sun or switch
Beamed from the pulse of an old awareness which
Tells me to age by love and not to cling
To ears, eyes, teeth, knees, hands—or any thing.

Elegy

Whenever my father was left with nothing to do—
   waiting for someone to 'get ready',
or facing the gap between graduate seminars
   and dull after-suppers in his study
grading papers or writing a review—
   he played the piano.

I think of him packing his lifespan
   carefully, like a good leather briefcase,
each irritating chore wrapped in floating passages
   for the left hand and right hand
by Chopin or difficult Schumann;
   nothing inside it ever rattled loose.

Not rationalism, though you could cut your tongue
   on the blade of his reasonable logic.
Only at the piano did he become
   the bowed, reverent, wholly absorbed Romantic.
The theme of his heroic, unfinished piano sonata
   could have been Brahms.

Boredom, or what he disapproved of as
   'sitting around with your mouth open'
oddly pursued him. He had small stamina.
   Whenever he succumbed to bouts of winter bronchitis
the house sank a little into its snowed-up garden,
   missing its musical swim-bladder.

None of this suggests how natural he was.
   For years I thought fathers played the piano
just as dogs barked and babies grew.
   We children ran in and out of the house,
taking for granted that the 'Trout' or E flat Major Impromptu
   would be rippling around us.

5

For him, I think, playing was solo flying, a bliss
    of removal, of being alone.
Not happily always; never an escape,
    for he was affectionate, and the household hum
he pretended to find trivial or ridiculous
    daily sustained him.

When he talked about music, it was never
    of the *lachrimae rerum*
that trembled from his drawn-out phrasing
    as raindrops phrase themselves along a wire;
no, he defended movable doh or explained the amazing
    physics of the octave.

We'd come in from school and find him
    crossed-legged on the jungle of the floor,
guts from one of his Steinways strewn about him.
    He always got the pieces back in place.   .
I remember the yellow covers of Schirmer's Editions
    and the bound Peters Editions in the bookcase.

When he defected to the 'cello in later years
    Grandmother, *in excrucio*, mildly exclaimed,
'Wasn't it lovely when Steve liked to play the piano.'
    Now I'm the grandmother listening to Steve at the piano.
Lightly, in strains from the Brahms-Haydn variations,
    his audible image returns to my humming ears.

What I miss

is some hexagonal white seal
like a honeycell.
Silence I miss;
the hand on the fiddle
muting the vaulted arrogance
it raises;
the crowded hush
of the conductor's lifted wand,
then the chorale
walking with little empty breaths
through air it praises.

My air is noises
amplified by an ugly pink
barnacle in my ear.
All the music I hear
is a tide dragging pebbles
to and away in my brain.
Sphered, the harmonies fall,
mutate, abort. Emptiness
is like rain
in my insomniac city,
ceaseless and merciless.

For Elon Salmon

'All Canal Boat Cruises Start Here'

A musk of kittening
snuffed out her morning dream—
the childhood tabby whelping on her mother's shoes;
still no one could find those
x-ray photographs of embryos.

'I didn't know I could dream smells,'
she said at breakfast.
'You're pregnant,' he said, 'you're a miracle.'
'I'm guilty where there's too much evil.'

Later they walked by the canal—
the Sunday crush, the peacefulness,
a crowd of sleepy explosives spoiling the waterfowl,
exclusive wagging dogs, lovers, their binary excuse,
the straggling families.

'They don't see us,
they're all wound up in themselves,' as she
looked understanding at one impervious mamma,
a daughter's long pale silk hair shining,
a sulky boy scuffing behind.

Was that it? That look flashing (was he?)
furtive, from a second-hand suit?
Scary, the shabby solitary, hating you
and apologizing.

In the fenced off zoo an elephant
took a tiny white keeper for a trot,
and heavens, that barge cat
leaping three feet of slime to the quay.

But in painful labour
it was blue sky and fishermen she remembered,
each hopeful alone
in his nest of a basket and stool.

'They didn't catch fish,
but love, what a guard of honour'
as the waving drawbridge of rods
bowed the pleasure boats through.

Welsh Pastoral

For my grandson, 10 March, 1988

After it had rained all winter
April brought a chill East wind.
Somehow the sun got used to it
and stayed cheerful. In kind
(for Wales) well-mannered weather,
daffodils stretched and unfurled.
New lambs slipped out of their mothers
into a washed world.
Soon clumps of them could be seen
like quartz or distant cataracts
on hills not yet hospitable
but intensely green.

Driving from Llanfihangel on the A496
we stopped by a minor pasture where a shivery
minute-old calf had just introduced himself
to the terrible twentieth century.
He lay on the rough mud amazed, while
his mother licked off the salt envelope.
Behind her a rose red umbilicus
hanging down like a second tail
held and repelled us.

Of course we wanted to see
if the herd, closing in, could help.
But cows take their time over
rites of assent or hostility,
and we were, of course, in a hurry.
So we never saw the calf get up
from its grab-bag of legs to suck,
or knew how its mother worked free
of that long, necessary,
embarrassingly tough red rope.

Stone Fig

The young fig tree feels with its hands
along the white sunny wall
and at the end of August
produces seven fruits,
seven royal fists that will be
runny with seed, ripe
with a musky honey that rarefies sweetness.

For the sick woman in the bedroom
behind the all-day-drawn curtains
I set the fruit in the sun
by the kitchen window.
Seven brimming wineskins
and a flint from the garden
she must have collected with a smile,
for it looks exactly like a fig.

A stone fig,
a hard, smooth, comfort-in-the-hand
Platonic Idea of Fig.
I watch the others daily for readiness;
this one, and now that one.

As if they were the last of her feelings
the old lady gobbles the ripe figs.

Quickly, quickly, such greediness.
She eats them like anaesthetic.
Here's pith in her pale fingernails,
purple on the stubble of her chin.

Her legs are dry twigs. She can't
trust them to take her to the toilet,
then back again. Her skin is mottled
with overripeness I look away from.

She wants, she smiles, to sleep
and sleep and sleep. And then
to sleep again.

Inverkirkaig

For Dorothy and Jonathan Brett Young,
  September, 1986

Bloodshed cries *Ai Ai*
in the colour of rowan berries.
Birds will eat wounds all winter.

Flayed hills keep breathing
through the slashing—
waterfall, fighterplane, waterfall.

The salmon leaps and fails.
Drowned to his thigh,
the man praises courage with a hook.

Six men, twelve wings, and the sky cracks
successfully. The squadron
is out of hearing when it falls.

With its gentian eye, the loch
praises gleaming and burnishing.
The salmon leaps and fails.

The man goes on casting and reeling.
A late rain erases in pitiless mercy
home, story, arterial berry.

Icon

The scene they play
is the midwife's
without the midwife.

Blood, groans
have drained into the gold,
and all her pain

is inward and to be.

The child
is like a prophet
on her knee.

A Doctor of Science.

In joy
his forehead
flexes in its sphere.

His hand
that claws her face
catches her tear.

Branch Line

The train is two cars linking
Lincoln and Market Rasen.

Late May. Proof everywhere from
smudged sunny windows
that the Economic Community
is paying the farmers for rape.

How unembarrassed they are,
wanton patches the colour of heat
dropped like cheap tropical skirts
on the proper wolds.

The trees have almost completely
put on their clothes.
They sway in green crinolines,
new cool generous Eves.

White hawthorns, too,
do predictable cold unclenchings.

As the train parts green field from gold
a spray of peewits fans up in a bow wave.

Is that a factory out there
where the sea might be?
A lighthouse? A tall methane candle,
lethal if it were to go out?

The train slowly judders and halts
for no visible reason.

Rooks squabble in a maple.
A blackbird ferries an enormous worm to a nest.
Staring cows bend again to their munching.

Meaningless life, I'm reading in the TLS,
a nexus of competing purposes . . .

God is impossible.

Life is impossible.

But here it is.

Three Poems for Sylvia Plath

Nightmare, Daymoths

December, 1987

A glass jar rattles its split peas and pasta.
Those cysts look innocuous, but they weave
through the kernels, hatching into terrible insects.
Something's on the floor there,
buzzing like a swat wasp.
A belly like a moist rubber thimble
sucks and stings my finger. *Ach,*
my heel reduces it to sewage.

String the creatures up, then.
Hang them on the Christmas tree.

They glisten there like fish, now softly
lengthen into milliners' feathers.
See, they are only moths, paper moths or horses,
not even paper but the Paisley curtain
sifting ashy patterns from the winter light.

Order, they order, *order*.

The flame gropes for a fire.
The dream asks meaning to patch its rags.
The flying words want paper to nest in.
Six colours rake the white reach of the rainbow.
Even the smallest hours crawl by with a number.

These letters are marching straight into an alphabet:
X Y Z, not to infinity.

Letter to Sylvia Plath

Grantchester, May, 1988

They are great healers, English springs.
You loved their delicate colourings—
sequential yellows, eggshell blues—
not pigments you preferred to use,
lady of pallors and foetal jars
and surgical interiors.
But wasn't it warmth you wanted most?

These Grantchester willows keep your ghost,
young and in love and half way through
the half-life that was left to you.
The Cam still crawls through patient grass,
preserving ephemerals in glass.
A bull thrush shouts from a willow thicket,
Catch it! Catch it! Catch it! Catch it!
Catch what? An owl in a petalled dress?
The gnarl at the root of a distress?

Dear Sylvia, we must close our book.
Three springs you've perched like a black rook
between sweet weather and my mind.
At last I have to seem unkind
and exorcize my awkward awe.
My shoulder doesn't like your claw.

Yet first, forgiveness. Let me shake
some echoes from old ballad eyed Blake
over your grave and praise in rhyme
the fiercest poet of our time—
you with your outsized gift for joy
who did the winged life destroy,
and bought with death a mammoth name
to set in the cold museum of fame.

Your art was darkness. No, your art
was a gulping candle in the dark.
In the beginning was a curse:
a hag, a drowned man and a nurse
hid in the mirror of the moon
unquietly to work your doom.
A dissolute nun, you had to serve
the demon muse who peeled your nerve
and fuelled your energy with hate.
Malevolent will-power made you great,
while round you in the Sacred Wood
tall archetypal statues stood
rooted in air and in your mind.
The proud impossibles loomed behind,
pilasters buttressing a frieze
of marble, moonlit amputees.

Sylvia, I see you in this view
of glassy absolutes where you,
a frantic Alice, trip on snares,
crumple and drown in your own tears.
You were your cave of crippled dreams
and ineradicable screams,
and you were the pure gold honey bee
prisoned in poisonous jealousy.

The gratitude and love you thought
the world would give you if you fought
for all your tears could not be found
in reputation's building ground.
O give the mole an eagle's soul
and watch it battling in its hole.

Because you were selfish and sad and died,
we have grown up on the other side
of a famous girl you didn't know.
The future is where the dead go
in rage, bewilderment and pain
to make and magnify their name.

Meanwhile, the continuous present casts
longer reflections on the past.
Nothing has changed much. Famine, war
fatten your Spider as before.
Your hospital of bleeding parts
devours its haul of human hearts,
excreting what it cannot use
as celluloid or paper news;
eye for eye and tooth for tooth,
bomb for bomb and youth for youth.

Yet who would believe the colour green
had so many ways of being green?
In England, still, your poet's spring
arrives, unravelling everything.
A yellowhammer in the gorse
creates each minute's universe;
a blackbird singing from a thorn
is all the joy of being reborn.

Even in Heptonstall in May
the wind invites itself away,
leaving black stone to compromise
with stitchwort, dandelions and flies.
Tell me, do all those weeds and trees
strewing their cool longevities
over the garden of your bed
have time for you, now you are dead?

Behind the pricked-out drape of night
is there a sheet-white screen of light
where death meets birth to reconcile
the contradictions of your will?
Perfection is terrible, you said.
The perfect are barren, like the dead.

Yet life, more terrible, maunches on,
as blood-red light loops back at dawn,
seizing, devouring, giving birth
to the mass atrocity of the earth.
Poor Sylvia, could you not have been
a little smaller than a queen—
a river, not a tidal wave
engulfing all you tried to save?

Rather than not be justified
you sickened in loneliness and died,
while we live on in messy lives,
rueful or tired or barely wise.
Ageing, we labour to exist.
Beyond existence, nothing is.
Out of this world there is no source
of yellower rape or golder gorse,
nor in the galaxy higher place,
I think, for human mind or face.

We learn to be human when we kneel
to imagination, which is real
long after reality is dead
and history has put its bones to bed.
Sylvia, you have won at last,
embodying the living past,
catching the anguish of your age
in accents of a private rage.

Hot Wind, Hard Rain

August 1988

The joy of the rowan is to redden.
The foxglove achieves the violence of its climb.
This summer gale flattens the flower
    and deforms the tree.
The dog trots at a queerer angle
    to the disused railway.
The tabby seizes the fledgling blown to the midden.
From the river, gaseous with weed, a reek of decay.

Hot winds bring on hard rain, and here in Durham
    a downpour tonight will probably allay
whatever has got the willows by the hair,
shoving light under their leaves
    like an indecent surgeon.
Now light's in every particle of air,
acetylene wind that blows too hard and clear.
Who sifts the saving from the killing terrors,
    O my dear?

Calendar

The blank days
are heaped up ahead of me,
a sierra I have to cross
in order to get there. Where?

The marked days
are a slag of words.
In the course of crossing
I must have upturned them. How?

The days for crossing
have a solid, permanent appearance—
mountains of sandstone and sun
between shadowy valleys of conifers.

The days I have crossed?
Every one of them misty rubble.
A penline suggests
there is hardness there. Or art.

And this day I am crossing,
crossing in sun and rain?

Seems perfectly flat.

The sun celebrates my defeat
as I struggle towards it.
The rain lays a wreath on each drop
as it dies in its puddle.

North Easter

April, 1986

That daffodil trumpets its *gaudia*
straight into the ground, whence
rhubarb arises—red knuckles,
green gleaming rucked sleeves.

But the wind wants everything
to be level with itself:
the crocuses, the washing on the line,
the soiled plastic streamers,
desperate to be blossoms,
impaled on thorns, nailed
to the skeleton trees.

Also rags from the gangrened top—
half of a patched cardboard suitcase;
also bright kelpy tanglements of
orange-tinted baler twine, kinking,
unkinking on the wind's dry littoral;
comb of tarred roofing-felt, rusty cans,
torn carpet, coat, boot, shoe.

What god will arise and slouch
through this realm of rubbish?
The low flower, the coltsfoot,
creeps in its silver scales between
glittering fragments of beer bottle;
the goat willows kneel, fists budding,
among no longer useful speedometers;
rising larks strew nervous hallelujas
in the cooked paths of the Yamaha.

There are no fish now in the river
squamous with air. Which is re-writing,
anyway, its book of revelations in the blood
of old sidings and carburettors.
A glory of oily rainbows stains the stones,
jewels also five drowned black tyres.

Night Walking with Shadows

Nightwalking the dog through the hollow village,
I am followed and preceded by three of me.

The streetlights distribute me between three shamans.
Their huge imaginations hand me, like a trophy,
From the shadow behind me to the shadow before me,

while the full moon gives me a dense
practical shadow, smaller than myself.

I walk, for the dog's sake, out of the lights
up the track by the sportsground, the shacky allotments.

How this white fall of moonlight simplifies the story.
Dog and shadow. Woman and shadow.

Up the V of the valley, a string of brilliants.
In every window, labouring magi.

The chimney pots steam like alembics,
but for every white chain of amazing smoke
the moon cuts a dead black track.

The Morden Angel

A monologue of the plaster bust of an angel who presides over poetry readings in the Morden Tower, Newcastle.

My sideways smile
means I'm wearing
the joke of my being
like a punishment.

The child of baroque
imagination, I could have
risen. In cloudy
theatrical surroundings,
say, with a pipe
on a ceiling, or burdened
with a wreath of pineapples,
my squint might not
have disadvantaged me.

But here, where the spirit of
Art, one winter,
snagged the career
of my creator, I came
into my little kingdom
crooked.

If it was a muse he needed,
why did my maker
make me dumb?
I am full to the lips
with iambic pentameters.
In couplets I might have
inspired him.

What has come
to my attention as 'verse'
in these late days
he wouldn't have believed.

Especially my wings
are worrying. Ought they
to wave like an angel's
or flutter like Cupid's?
They pin me to the wall
like a target.

It's from trying to
fly away from this wall
or back into the wall
that my temper suffers.

Here come those poets again,
their inelegant wails!
If only my wings would work.
If only my smile could talk.

And even then,

there may be a language in which
memory will be called 'letting in the sorrow'.
It would be a black language.
The sorrow would be a rainbow
after the storm, at its beginning.

Music in this language would mean
'measuring the rhythms', and poetry,
'translating the dreams'.
Power? A hush in which to honour
winds' work, and the sun's.

A long litany of astonishment
would be, in this language,
a hymn of thanksgiving: 'Even as it died,
the sea made power out of its own pulse,
pounding to salt the poisoned cities
of the suicides.'

Call them Poppies

Imagine a reconciliation between
dumb eyes and their tears.
Full pools watch it all without blinking.
Ploughed dust, brick rubble and blood
Accumulate on the lashes.

Now the terrible red wreaths
Glow like coals. Call them poppies.
It's when each held-back tear
Breaks its brim like a waterfall
That the mouth cries the shape of its store.

The Mass, The Media, The Market

The Mass

The Hydra in his
   ruff of heads would not
have believed it—
   each of my necks is
attached to a different name.

Our appearance as one is
   achieved by the
tensions between us,
   a balance of tentacles
touching the
   green of this earth.

Great appetite and suction
   contract us
advancingly together. Why
   should we war?
We are one growth. What I
   consume you shall
consume. More must never
   be enough.

The Media

The descent into Hades
   will cost you three senses.
Take off your touch,
   please, your noses and mouths,
you won't be needing those.
   Eyes, though, and ears,
with their nerves, you

may preserve. Shall I
inflate them for you?
    You'll find it smoother,
really, if you just let your
    bodies drop off casually.
There! Didn't I
    tell you? Your heads will
hold you up. You see
    how nice it is,
floating in the other
    life. No trouble to
travel; suffer when you
    wish, without personal
pain. It's all literature
    with no books, you
score with the mighty, you
    hear the gods speak.
Why ever go back to the
    feeble squalor of bodies?
So little is possible where
    anything's real.

The Market

I am your child.
    You may call me
Midas. I rule because you
    make the rules. Your
health means my
    growth, so I keep faith
touchingly. Now you're
    crawling for miles
towards my gorgeous
    affliction. I
offer you my mouth, my
    resplendent intestine.
You feed me your

time, your properties,
your investments.
   It's all in your
interest I've
   swollen to this size.
Though you love me it's
   time for you to be
careful. I'm bigger than
   three-hundred million of you
swarming together. I'm
   the one baby
in the world who chews
   flesh into cash.

Inquit Deus

The world is the world
but you ask it the wrong questions:
    What can I make of it?
    When will it pay me?

Its full stop weighs in your palm
like a pitted moon.
    Why do I suffer?
    When will you save me?

Here's your life line, vertical,
your head line, horizontal,
your fingers off in space
with that galaxy, your mind.

And this sea-bitten pebble
you're so perilously keeping
is all you have to live for,
and its love is blind.

From the Motorway

Everywhere up and down the island
Britain is mending her desert;
marvellous we exclaim as we fly on it,
tying the country in a parcel,
London to Edinburgh, Birmingham to Cardiff,
No time to examine the contents,

thank you, but consider the bliss of
sitting absolutely numbed to your
nulled mind, music when you want it,
while identical miles thunder under you,
the same spot coming and going
seventy, eighty times a minute,

till you're there, wherever there
is, ready to be someone in
Liverpool, Leeds, Manchester,
they're all the same to the road,
which loves itself, which nonetheless
here and there hands you trailing

necklaces of fumes in which to be
one squeezed breather among
rich and ragged, sprinter and staggerer,
a status parade for Major Roadworks
toiling in his red-trimmed triangle,
then a regiment of wounded orange witches

defending a shamelessly naked
(rarely a stitch of work on her)
captive free lane,
while the inchlings inch on
without bite or sup, at most
a hard shoulder to creep on,

while there, on all sides,
lie your unwrapped destinations,
lanes trickling off into childhood
or anonymity, apple-scented villages
asleep in their promise of being
nowhere anyone would like to get to.

From the Primrose Path

Effulgence or copper polyphony
where once one cough from the Lord
produced, in a vermilion corona,
Gabriel with a fiery sword
leaping from the hot heart of punishment.
Now it spills over Regent's Park
a phenomenon of city-vapour, light,
and a catalyst of human perception,
smelting for this rare, Marylebone ceiling
a luminous ore.

How well Nash's domes suit the smooth-bellied
mosque and Planetarium, those Muslim minarets,
our narrow Victorian steeples;
even the Telephone eyesore
gets on in a flat silhouette, leaning
on a spectacle nobody thinks to pay for—
as tonight, on the footpath to Baker Street,
where a flash of high baroque evidence
checks us, I noting chill in the wind, you,
Concorde, manned serpent gilding in azur,
eyeing the continents.

Night Thoughts and False Confessions

He:

How uneasily I live
in the house of imagination.
True beams drive cleanly, sliding,
rising from hissing traffic, only to
slip or miss on your chancy ceiling.
We lie in your bed's dark negative.
I smell you not sleeping.
I can feel you being human and woman.
On the slopes of the absolute
absolutely nothing would happen.
The world would revolve and evolve
pure white, stone-white and blue,
the beautiful figure of the desert,
musical and mathematical, its deep throat
blameless, immaculate, swallowing you.

She:

With wary brutality
the thrush cracks the house of the snail.
When you said you were leaving me
your face flashed beaky and cruel.
My shell disintegrated completely.
You beat me and beat me
on the slab of your mind's concrete.
I was sweetness in your kitchen,
obedience in your word-stall.
I was softness overflowing in your garden
scented deep at nightfall.
I cry for lost days, for youth,
for love questioned too late
as the cooked trout weeps white eyes
on the gourmet's plate.

Journal Entry: Impromptu in C Minor

Edinburgh, October, 1988

After weeks of October drench,
a warm orange day,
a conflagration of all the trees and streets in Edinburgh.

Let me have no thoughts
in this weather of pure sensation.

Getting into the car is a coatless sensation.
Driving through the traffic
is the feeling of falling leaves.

The Firth, like the sky, is blue, blue,
with sandy brown puffs of surf on the oily beaches.
The sea swell rises and spills,
rises and spills, tumbling its load of crockery
without breakage.

Is a metaphor a thought?
Then let these shells be shells,
those sharp white sails be sails.

Today the pink enormous railway bridge,
is neither a three-humped camel nor a dinosaur
but a grand feat of Scottish engineering;
now and then it rumbles peacefully
as a tiny train, rather embarrassed, scuttles across it.

Sitting with pure sensation on the breakwater,
I unhook the wires of my mind.
I undo the intellectual spider's web.
Uncomplicated me.

But I correct myself.
Soon I'm standing in my grid of guilts
hastily reaching for my thoughts.

For there are people out there.
Not abstractions, not ideas, but people.
In the black, beyond the blue of my perception,
in the huge vault where the wires won't reach,
the dead are lively.
The moment I take off my thought-clothes
I expose every nerve to their waves.

What is this sad marching melody?
A spy, a column on reconnaissance,
the theme from Schubert's Impromptu in C minor.

It is 1943.
In a frame house that has forgotten him,
a dead man is playing the piano.
I am ten years old. For the first time
I watch a grown woman weep.
Her husband, the white-haired Jewish philosopher,
makes shy mistakes in English.
He puts an arm around his wife
and bows his head.

The theme returns years later
to a farmhouse in Vermont.
This time I myself am at the piano,
a puzzled girl I instantly recognize
although she died through more years than Schubert lived
to make room for the woman I am now.

I smile at her ambition.
She doesn't yet know she will be deaf.

She doesn't yet know how deaf she's been.

What is the matter?

40

This is the matter: deafness and deadness.
The shoe-heaps, hills of fillings, children's bones.
Headlines blacking out the breakfast chatter;
　　(We go on eating).
Static and foreign voices on the radio;
　　(We are late for school).

Then silence folding us in,
folding them under.

But here is the melody.

And here, 'our daemonic century'
in which a dead man's dead march
plays itself over and over
on a fine fall day in South Queensferry
in the head of a fortunate (though deaf) American
　　grandmother

She sits in the momentary sun looking at the sea.

Once there lived in Austria a schoolmaster's son,
shy, myopic, a little stout, but lucky,
for his talent was exactly suited to his time.
Careless of his health in an age of medical ignorance,
he died at thirty-one, probably of syphilis.
A few moments of his life, five notes of it,
fuse with a few impromptu responses,
a few contemporary cells.

They provide the present and future
of an every-minute dying planet
with a helix, a hinge of survival.

Cramond

'Then 'twas the Roman, now 'tis I.'
               —Housman

Remember how in Edinburgh
on fine spring evenings, the family cars
shook out their dogs and children
to animate the live Scots postcard
at Cramond?

Seaward spread the too blue Firth,
patronized by sandpipers.
A far fringe of mountains
framed a classical view.
You could walk to

Cramond Island when the low tide
let you. But mostly you wanted to
skip in your mind from peak
to concrete peak across the buckling
ship blockade.

Somehow that was simpler than a
free-fall to the Romans, their mute frontier
devoid of sirens and motorbikes
and slow planes homing over the oil fields
like metal pigeons.

Celebrity

When I lie down to sleep
with my reputation
serving the name of my dust
like a grinning doll,
I'll no longer need to remember
any occasion
of human indignity or fuss
to explain it all.

No ill-concealed file of my faults
will ever extend this
crotchety itch of becoming
I know as me
beyond the illustrious face
an approving, vicious
future of sorting and smoothing
decides to see.

As in tennis, love means nothing;
my famous matches
are there for the faithful
in lights that stud the court;
with an inrushing wave of applause
that in praising, passes,
along with the cash, and my skilful
wasted art.

Eros

I call for love
But, help me, who arrives?
This thug with broken nose
And squinty eyes.
'Eros, my bully boy,
Can this be you,
With boxer lips
And patchy wings askew?'

'Madam', cries Eros,
'Know the brute you see
Is what long over use
Has made of me.
My face that so offends you
Is the sum
Of blows your lust delivered
One by one.

We slaves who are immortal
Gloss your fate,
And are the archetypes
That you create.
Better my battered visage,
Bruised but hot,
Than love dissolved in loss
Or left to rot.'

Seven Poems after Francis Bacon

1 *Study for a Portrait on a Folding Bed*

Squirmed flesh on the blue folding bed,
ear torn from a head,
he's the pupil of his eye
at the heart of his house of glass.

Afloat in his well, being, why
does his black palette drip
from that lap of aggression? What
nightmare will the egg hatch
that lies—a prediction of
greenlessness—on that dead step?

2 *Study of a Dog*

The last dog
rises and tiptoes out
of the peagreen hoop which is
all that's left of the park.

It's hot on the penthouse roof,
which is all that's left of Monte Carlo.

The shade of the last tree,
on the far side of the motorway.

Is it water he wants,
or that little cross ball
rolled out of the sand trap
onto the burning turf?

3 Three Figures and Portrait

Personality, the flesh artist
(and presiding self-portrait),
has perfected his trick of exposing
bodies to money until
they attain metallic putrefaction.

When these hunks of human stuff
harden into fixed forms,
reptilian man schemes in one circle,
mammalian woman in another.

Her gold ring is larger and brighter
than his rink of empire.

That fury in a cage
is their fortunate offspring.

She has sprung fully armed
from their heads,
and she means to hate them.

At present she's
spooling her childhood
in her chittering skull.

4 Seated Figure

Self-placed,
for years he's struggled to achieve it,
a position exactly at the centre
of his invisible frame.

The expensive Kirman
congratulates his shoes.

The rich blue furniture, discreet blue curtain,
push forward his expression of
resourceful authority, a farsighted
gaze to the future, though

tensely he holds in readiness
shortsighted spectacles, for details.

You can put your trust in me,
says the strong right shoulder,
but the left will not bear
the weight of responsibility.
He's had to remove it
to centre the picture on
righteousness.

Shift him three inches to the left
and everything's lost.

5 *Portrait of a Lady*

'I did, of course, consent to it.
I've never been interested in flattery.
Character is your forte, Mrs Allstone,
(dear Motherwell, I think, or was it Hockney?)
Sir Geoffrey warned me in Paris,
You won't want to hang it in your boudoir.
No way, I told him, my scene's
the sex war, not the boud war.

Now look at my right eye.
There I am, you see, my taste, my power.
The mouth is a betrayal.
I've been abused, you know,
for my generosity. I have risked.

I have guessed. I have made good
my investments. You will find my name
among the Friends of the major galleries.
My calling's to be a friend of artists.

That I was wiser than my husband
with money was *not* the reason
for our separation. Shall we say
we differed in numerous big ways?
I think the misery of my mouth pays
for the height of my brow.'

6 Triptych

In hades
by the swimming pool,
seated, he dries
right foot with
right hand.
Torso has slipped
to the floor,
rosy towel.

Drift flesh
copulates with
old imperative
desire.
Womb lies beneath them,
pink bib, flat pad.

Nearly whole
& fixed
& talking,

this experiment
(fleshing out
adipocerous soul)
has almost
succeeded.

7 *Study for a Portrait of Van Gogh*

I paint the rich and the damned.
They wince and buy me.
He only painted the sun
as poor as himself.

I live in the world of the world,
its rings of interior.
He lived in the circle of his hat
and through its gold weave
gazed at the angels
like grasshoppers in the heat.

I see the white sweep of his wings
but I paint white paint.

I lay the whole weight of my gift
on his stooped back.

Journal Entry: Ward's Island

For Lauris Edmond
Toronto, February, 1989

On the last day of the poetry festival
I took the harbour ferry to the island.
Two dollars to crawl across the iced silk
of my view from the hotel window
from the giant stalagmites of the waterfront
to Ward's back woods.

A sunny Sunday, cobalt, with feathery clouds.
Wind at minus eighteen degrees centigrade
knifing the lake. I felt on my face
the grinding of its blade. But water tossed light
like fish scales in our wake, so I stayed on deck
in my boots and visible breath
watching the city recede, its rich brocade
silver, jet, gold, porphyry, jade.

Splendid in the sun, with their mineral eyelids
the tall banks winked goodbye from a haze of piers.
After fifteen, twenty-five minutes
that could have been fifty years,
with a backwash of waves and a shrill
admonitory hoot, fussily we arrived.
Someone's sleeve threw a rope,
someone's caked gloves tied it.
I skidded down a gangway from the boat
to where Ward's little wilderness survives.

Outcast, silent, down-at-heels, deserted.
Whose shacks of rotting clapboard, synthetic brick?
Whose mud yards? Whose buggies and broken toys?
  Whose chairs
left out for a snow picnic, maybe?

Who stacked the wood so neatly? Who tarred the roofs?
Whose smoke? Whose cat? Whose bandaged porch? Whose
   story?

Talking to myself. In that place even the cat
seemed disposed to be sad and scarce.
Perhaps he was old Ward's ghost
scuttling by in the fur of his past.
I heard my footsteps one by one talk back
along a windswept municipal boardwalk:
evergreens, benches, tourist views of the lake.

On the other side, the open side, colder.
I saw how ice had hugged and hugged each boulder;
the beach was studded with layered, glittering skulls.
I kept on walking, with whatever it was I felt—
something between jubilation and fear.
There appeared to be no traffic at all
on that sea no one could see over;
only to the airport frail, silver insects
sailed from the beautiful air.

Turning, I cut through forest to the canal.
Two boys in red and blue padded anoraks
skated swiftly between frozen-in boats.
They'd set up rusty oil cans for goal posts,
and the faint flick, click of their hockey sticks
knitted me a coat.

Still, I caught the next ferry back.
A gaunt youth in a baseball cap and two burly men
settled themselves and their boredom
in the too hot cabin,
there to spread newsprint wings and disappear.
I paced the warmth, rubbing life into my hands.
The city advanced to admit us, cruel and dear.

Little Paul and the Sea

Hi yi hee yippee!
It's little Paul
dancing for the sea.

He's so small
the sea won't look at him
at all,

rising up
in a net-white heap,
falling dead
on the sand at his feet.

Such a huge blue reach
to the sky,
and every wave
rolling in to die.

Which is not what appears
to Paul,
for whom it's a nice
big pool.

Hi yi hee yippee!
It's little Paul
stamping on the sea.

OXFORD POETS

Fleur Adcock

James Berry

Edward Kamau Brathwaite

Joseph Brodsky

Michael Donaghy

D.J. Enright

Roy Fisher

David Gascoyne

David Harsent

Anthony Hecht

Zbigniew Herbert

Thomas Kinsella

Brad Leithauser

Derek Mahon

Medbh McGuckian

James Merrill

John Montague

Peter Porter

Craig Raine

Christopher Reid

Stephen Romer

Carole Satyamurti

Peter Scupham

Penelope Shuttle

Louis Simpson

Anne Stevenson

George Szirtes

Grete Tartler

Anthony Thwaite

Charles Tomlinson

Chris Wallace-Crabbe

Hugo Williams

also

Basil Bunting

W.H. Davies

Keith Douglas

Ivor Gurney

Edward Thomas